ROCKS & MINERALS

FOSSILS

BLASTOFF! DISCOVERY

BY PATRICK PERISH

BELLWETHER MEDIA • MINNEAPOLIS, MN

Blastoff! Discovery launches a new mission: reading to learn. Filled with facts and features, each book offers you an exciting new world to explore!

This edition first published in 2020 by Bellwether Media, Inc.

Library of Congress Cataloging-in-Publication Data

Names: Perish, Patrick, author.
Title: Fossils / by Patrick Perish.
Description: Minneapolis, MN : Bellwether Media, Inc., 2020. |
Series: Blastoff! Discovery: Rocks & Minerals |
Audience: Ages 7-13. | Audience: Grades 3 to 8. |
Includes bibliographical references and index.
Identifiers: LCCN 2019001754 (print) | LCCN 2019010052 (ebook) | ISBN 9781618916457 (ebook) |
ISBN 9781644870730 (hardcover : alk. paper) |
ISBN 9781618917409 (pbk. : alk. paper)
Subjects: LCSH: Fossils–Juvenile literature. |
Paleontology–Juvenile literature.
Classification: LCC QE721.2.F6 (ebook) | LCC QE721.2.F6 P47 2020 (print) | DDC 560–dc23
LC record available at https://lccn.loc.gov/2019001754

Editor: Betsy Rathburn Designer: Jeffrey Kollock

Printed in the United States of America, North Mankato, MN.

TABLE OF CONTENTS

DIGGING UP THE PAST

The summer sun beats down on South Dakota's endless **prairie**. The flat land is covered with tall grasses. A group of **paleontologists** carefully chips away at a rock. After hours of work, a fossil begins to emerge!

Other fossils are found nearby. The **impression** of a large leaf seems out of place in this treeless landscape. Thousands of years ago, this prairie was a lush jungle crawling with dinosaurs!

WHAT ARE FOSSILS?

Fossils are the preserved remains of ancient **organisms**. Fossils are at least 10,000 years old. Some are more than 3 billion years old! Paleontologists study fossils to understand ancient life. They dig up fossils from rock layers.

In most cases, fossils are not the actual bodies of organisms. Instead, they are rocks or **cavities** left in rock. Soft materials such as hair do not preserve well. Hard materials like bones make better fossils!

PALEONTOLOGIST EXCAVATING

HIGH SPEED CHASE

Fossilized footprints in Texas bring to life an ancient chase. Sauropod tracks are closely followed by a *T. rex* relative!

Fossils are not just the remains of dead organisms. **Trace fossils** record the activity of ancient plants and animals. These include footprints, tooth marks, and burrows. For example, *Asteriacite* fossils are markings left by sea stars digging into the sand!

Sometimes, trace fossils are all that remain of soft-bodied animals. Some ancient worms are known only by their trails and droppings. Other times, trace fossils cannot be identified. More study may lead to more clues!

FOSSIL PROFILE

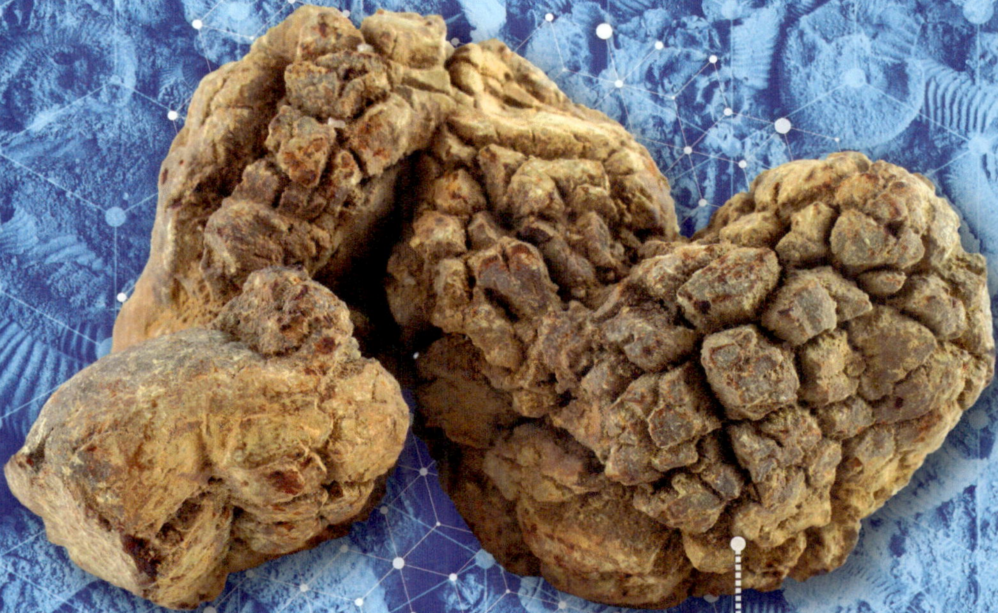

FOSSIL: **COPROLITE**

TYPE: trace fossil

FOUND: all over the world

MADE OF: fossilized poop hardened by minerals

USES: offers scientists information about what ancient animals ate

Body fossils are the rocky remains of organisms. Often, only a few bones are found during fossil digs. Complete skeletons are very rare. In larger finds, fossils may be jumbled together. Separating them can be a difficult job. It is like working on a puzzle with both missing and extra pieces.

Large body fossils are impressive. A dinosaur named Sue is the largest *Tyrannosaurus rex* ever discovered. Sue is also the most complete *T. rex* fossil ever found!

COMPARING FOSSILS

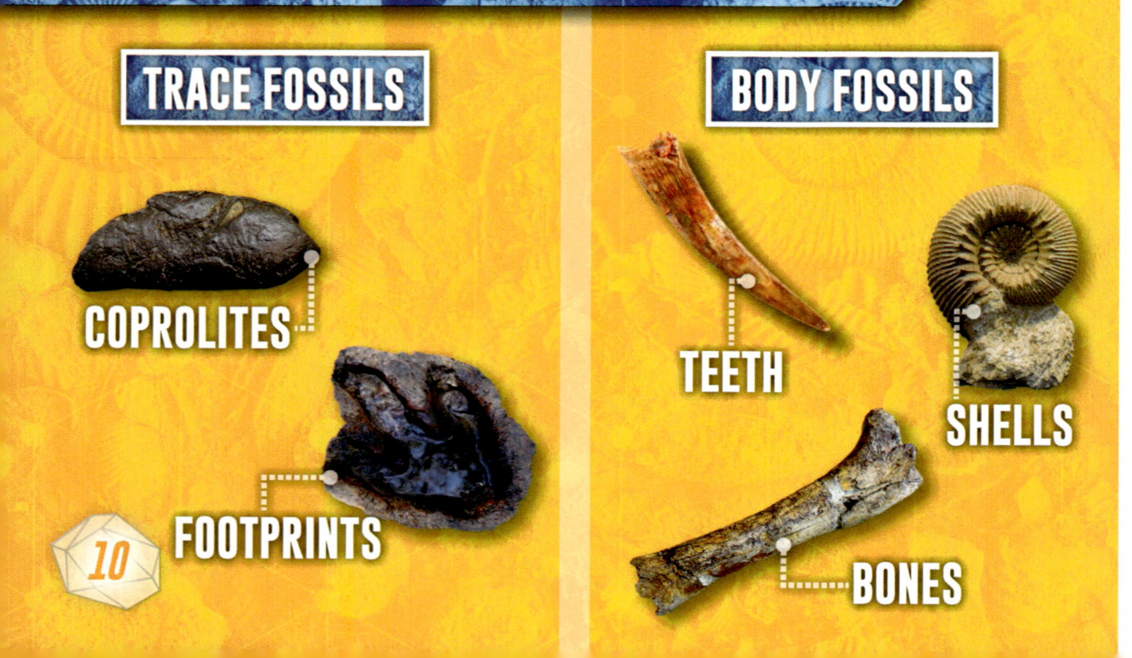

WHAT A FIND!

Sue was found in South Dakota in 1990. The dinosaur is named after the paleontologist who found her, Sue Hendrickson!

Most fossils are very small. Microfossils can only be seen with a **microscope**. These fossils are the preserved remains of bacteria, plants, and animals. Foraminifera are a very common microfossil. The fossils of these tiny, shelled organisms are found mostly in water.

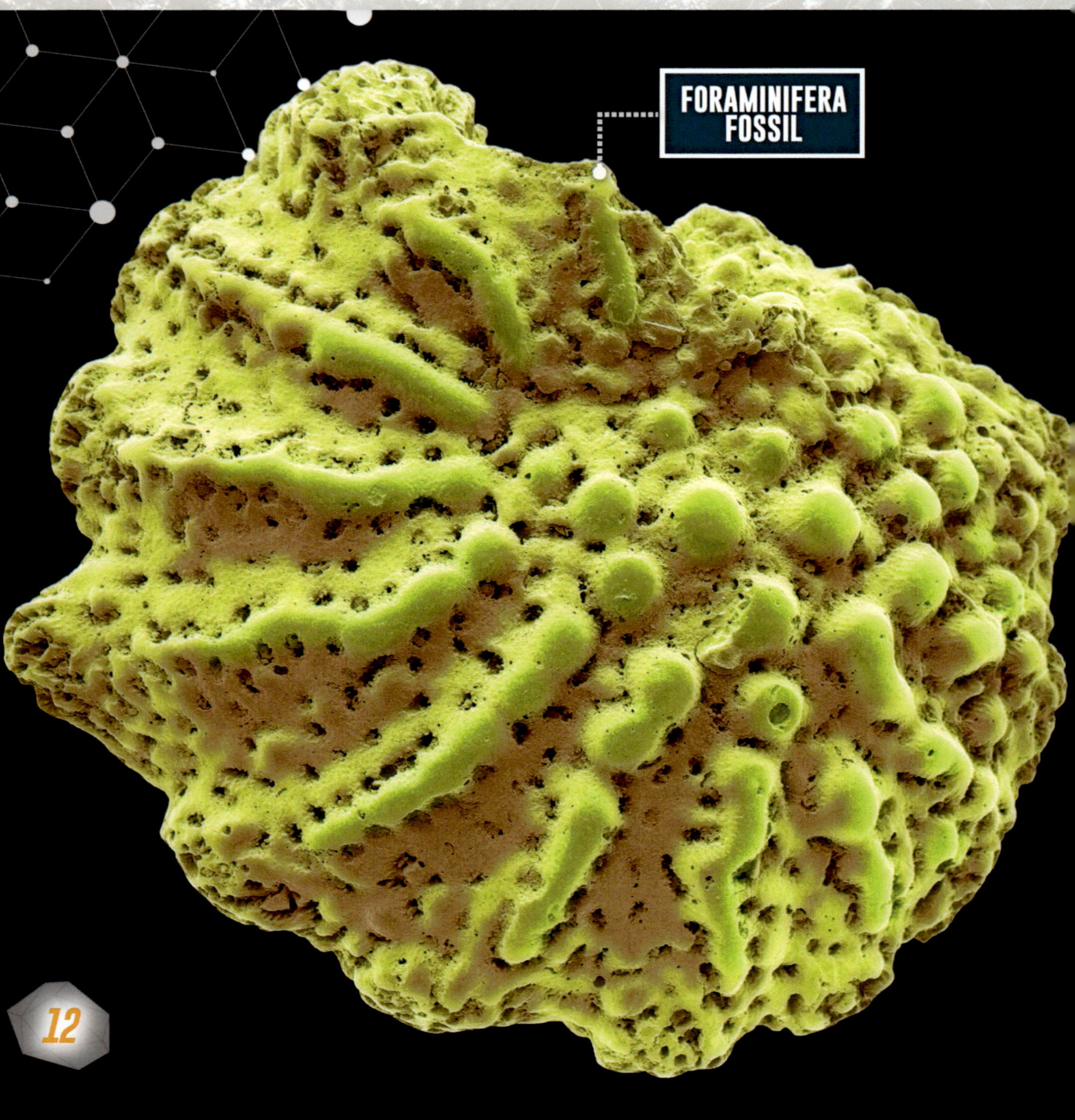

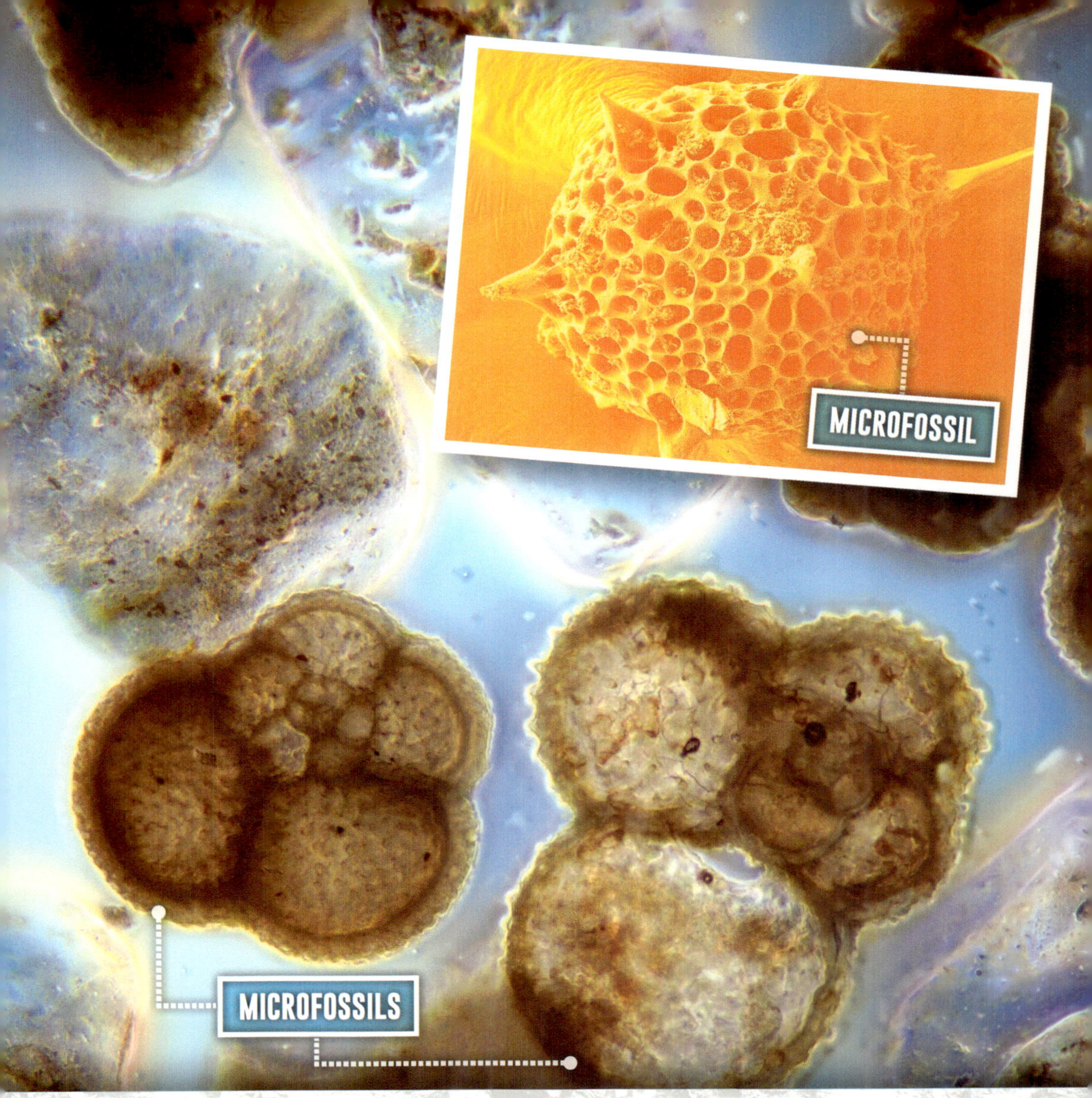

Microfossils offer scientists a better picture of early life on Earth. The number and types of microfossils found reveal how the atmosphere and oceans have changed.

HOW DO FOSSILS FORM?

Most organisms do not become fossils when they die. They are quickly broken down by insects and bacteria. Organisms must be preserved to become fossils. This can happen when they are quickly buried under **sediment**. Sediment prevents organisms from **decomposing** too quickly.

Quick burials are more likely in water than on land. Waves and powerful currents can shift sands. This is why the most common fossils are of water creatures.

LIVING FOSSILS

"Living fossils" are organisms that appear mostly unchanged from their fossilized relatives. For example, the coelacanth has not changed much for millions of years!

Fossils form in several ways. Hard fossils are preserved through **petrification**. After a bone or tree is buried in sediment, water seeps in. The water carries in **minerals**. These minerals fill in the gaps and harden the remains. The petrified forests of Arizona were formed this way!

Replacement is a similar way that fossils form. In this process, remains are completely worn away and replaced by minerals. The minerals preserve the shape of the remains. This process has created beautiful trilobite fossils made of shiny pyrite!

REPLACEMENT

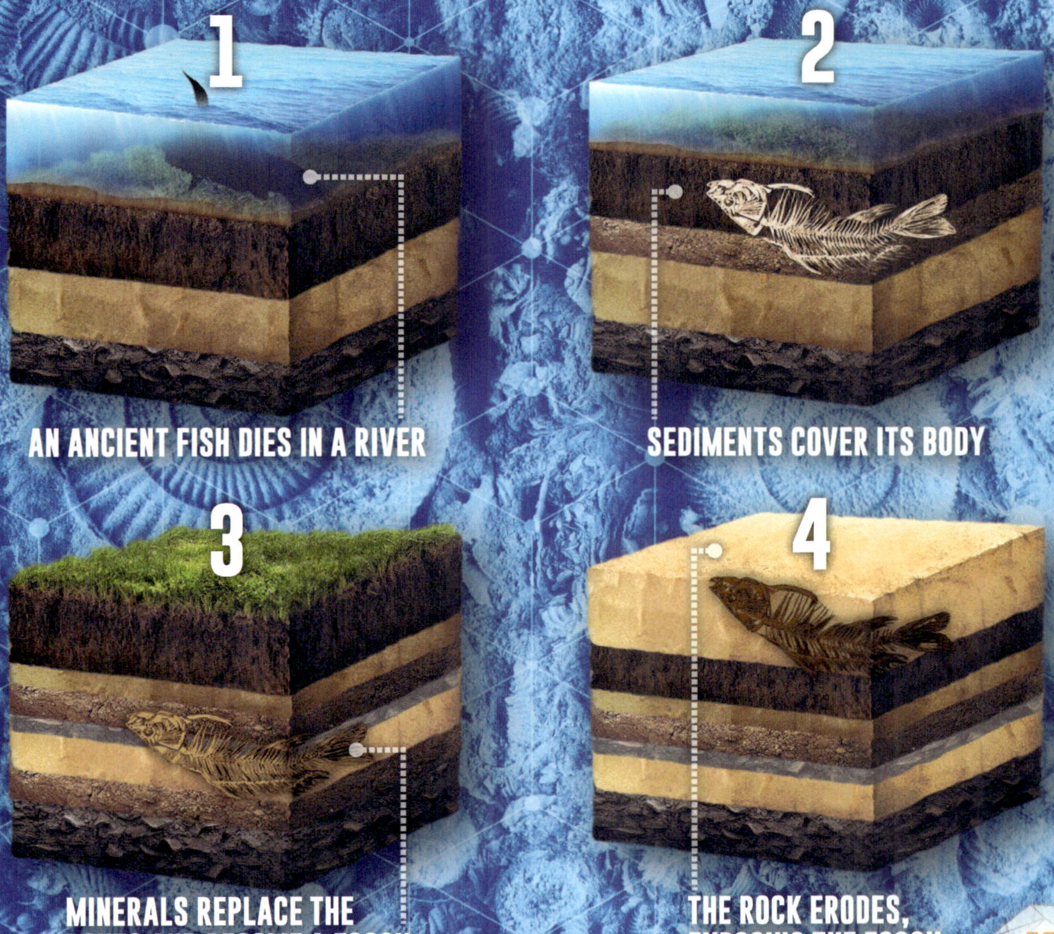

EARLY ANIMAL

The earliest known animal was *Dickinsonia*. It lived 558 million years ago. Many *Dickinsonia* casts have been found!

Molds are another type of fossil formation. Molds are impressions left behind in mud or dirt. They form after the remains of an organism are completely gone. The impression hardens into a fossil over time. Trace fossils such as footprints are formed this way.

Casts are sometimes formed from molds. Remains make a mold. The impression fills in with new material. The new material takes the shape of the original remains. It hardens into a fossil over time!

MOLDS AND CASTS

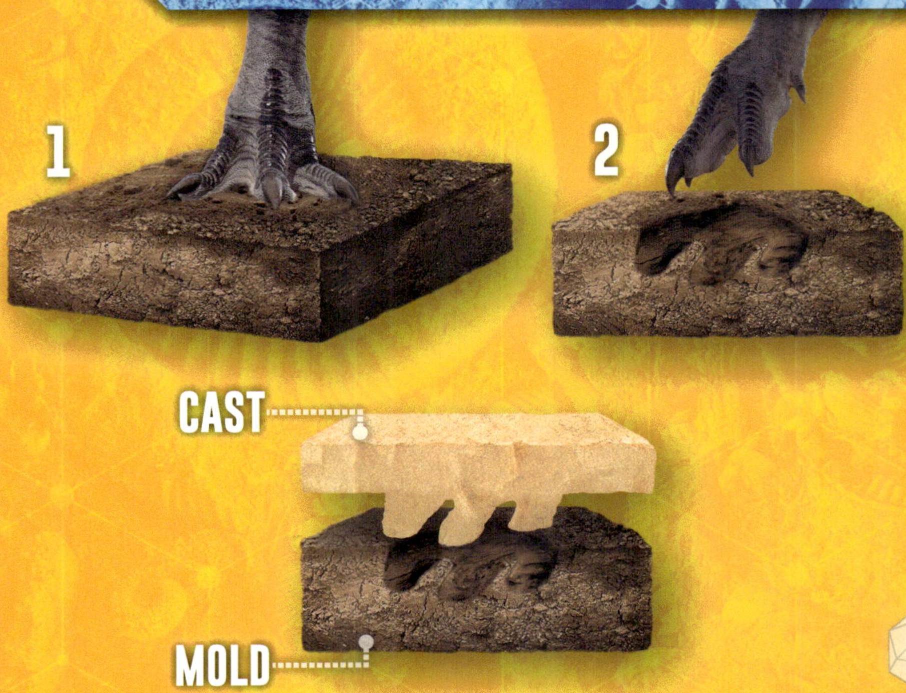

Soft organisms usually break down too easily to be preserved. But there are some exceptions to this rule. Some trees give off a sticky liquid called **resin**. Resin traps insects before it hardens. Then it gets mixed into sediment deposits. This fossilized resin is called amber. Amber preserves ancient insects!

TAR PITS

Bodies can also be preserved in tar. California's La Brea tar pits contain saber-toothed cats and other fossils of ancient animals.

In a few rare cases, organisms have been preserved in ice. A frozen mammoth was discovered in 2007 in Siberia. The 42,000-year-old baby was almost perfectly preserved. Its skin and hair were still intact!

Most fossils are found in specific areas around the world. Scientists look for clues in rock layers. Sediments that cover remains turn into sedimentary rocks. Exposed sedimentary rock layers are likely to contain fossils.

SEDIMENTS IN BELL ROCK, ARIZONA

Most fossil hunting occurs in deserts and other dry, rocky areas. There, exposed rock and few plants make fossil hunting easier. Fossils are often found in ancient riverbeds, bogs, and areas near volcanoes, too. The largest fossil discoveries have been made in the western United States, Argentina, and China.

HOW ARE FOSSILS USED?

Together, all fossils make up the **fossil record**. The fossil record has taught scientists a great deal about human life. It has helped in the study of **evolution**, a process in which living things change over time.

Fossils tell scientists how organisms have changed. Scientists look at fossils of the same type of organism. They study how the organism evolved over time. From this work, scientists have found that small dinosaurs evolved into today's birds!

DINOSAUR FOSSIL IN A MUSEUM IN FRANCE

Fossils are used in dating rock layers. Sediment is deposited in layers over millions of years. Organisms change over time through evolution and **extinction**. By matching fossils in rock layers with time periods, a rock's age can be determined.

Fossils also show how landscapes have changed. Giant shark fossils show that Kansas was once covered in ocean. Fossils of ocean creatures found in the Himalayas show how seafloor can be lifted into mountains!

MAKE YOUR OWN FOSSIL

MATERIALS

- 1 CUP SALT
- 2 CUPS FLOUR
- ¾ CUP WATER
- SMALL BOWL
- ANY SMALL OBJECT SUCH AS A SHELL, AN ACORN, OR A SMALL TOY

DIRECTIONS

1. Mix salt, flour, and water in small bowl, creating a soft dough.
2. Take a small section of the dough and flatten it.
3. Press a small object into the dough, making an impression.
4. Take the object out of the dough. You have made a fossil from a mold! Let the dough dry.
5. When the dough is dry, try filling the mold with more soft dough. Remove the dough, and you have made a fossil from a cast!

When organisms cannot change with their environment, they go extinct. Many organisms have gone extinct in the history of our planet. These extinctions are recorded in the fossil record.

The fossil record tells us how ancient climates changed and what the effects were. As the planet warmed or cooled, different species thrived. By studying fossils, we can better understand our changing world. They could be the key to keeping our planet healthy!

GLOSSARY

body fossils—fossils that preserve the bones, teeth, and other body parts of living things

casts—fossils formed from sediment filling a mold; fossil casts are the sediment that hardens over time.

cavities—holes or hollowed out pockets

decomposing—breaking down from being exposed to water, insects, or bacteria

evolution—the process in which plants and animals change slowly over long periods of time

extinction—an event in which an entire plant or animal species dies out

fossil record—the total collection of fossils that have been discovered

impression—a mark left by pressing a hard object onto a softer material

microscope—a tool used by scientists to look closely at rocks, minerals, or other materials

minerals—materials that make up sediments

molds—fossils that are formed when the remains of an organism decompose and leave behind an empty space in the shape of the organism

organisms—living things

paleontologists—scientists who study ancient life through fossils

petrification—the process of minerals from water filling the empty spaces within an organism's remains; the minerals harden to create a fossil.

prairie—a flat area of land covered in grasses

replacement—a way in which fossils are made where minerals completely replace an organism's remains and then harden over time

resin—a sticky liquid given off by some trees; insects are often fossilized in resin.

sediment—tiny pieces of dirt, rock, and other materials that become sedimentary rocks over time

trace fossils—fossils that preserve evidence of living things, such as their footprints or burrows

TO LEARN MORE

AT THE LIBRARY

Arnold, Caroline. *Living Fossils: Clues to the Past.* Watertown, Mass.: Charlesbridge, 2016.

Lessem, Don. *Ultimate Dinopedia: The Most Complete Dinosaur Reference Ever.* Washington, D.C.: National Geographic, 2017.

Taylor, Paul D. *Fossil.* New York, N.Y.: DK, 2017.

ON THE WEB

FACTSURFER

Factsurfer.com gives you a safe, fun way to find more information.

1. Go to www.factsurfer.com.
2. Enter "fossils" into the search box and click 🔍.
3. Select your book cover to see a list of related web sites.

INDEX

The images in this book are reproduced through the courtesy of: Ryan M. Bolton, front cover (tooth), p. 10 (tooth); Moha El-Jaw, front cover (trilobite); ntv, front cover (amber); Puwadol Jaturawutthichai, pp. 3, 8; Jason Patrick Ross, pp. 4, 5 (prairie); servickuz, p. 5 (fern); Cory Richards/ Getty Images, p. 5 (paleontologists); Marcio Jose Bastos Silva, p. 6; paleontologist natural, pp. 6, 7; Alex Coan, p. 9; Poozeum/ Wikipedia, p. 10 (coprolite); Rattana, p. 10 (footprint); MarcelClemens, p. 10 (shell); Zack Frank, p. 10 (bone); Jacopo Werther/ Wikipedia, pp. 10, 11; Science Photo Library/ Alamy, p. 12; Scenics & Science/ Alamy, p. 13; ClaudiaSEM, p. 13 (frame); Lukiyanova Natalia frenta, p. 14; Merlin74, pp. 14, 15; Frank Bach, p. 16; Universal Images Group North America LLC/ Alamy, p. 18 (Dickinsonia); APIWICH PUDSUMRAN, pp. 18, 19; Wasan Ritthawon, p. 20 (resin in frame); NitroCephal, p. 20; KAZUHIRO NOGI/ Getty Images, p. 21; Harry Beugelink, pp. 22, 23; Jordon Sharp, p. 23 (fossil rock); SAPhotog, p. 23 (layered rock); benedek, p. 24; LUDOVIC MARIN/ Getty Images, pp. 24, 25; Duard van der Westhuizen, p. 26; Russ Vance, pp. 28, 29; Kit Leong, p. 29; Gilda Villarreal, p. 31.